How to Watch Movies and Television Shows For Free

Other Titles by Annie Jean Brewer:

Nonfiction Titles:
The Shoestring Girl: How I Live on Practically Nothing...And YOU Can Too!
How to Start Out or Over on a Shoestring
The Minimalist Cleaning Method
400 Ways to Save a Fortune
How to Write and Sell an Ebook
How to Write Ebooks For a Living
Where to Work Online
Professional Help: How to Prevent and Fix Malware, Viruses, Spyware and Other Baddies
How to Watch Stuff Online For Free
Be Happy Now
How to Be Happy

Fiction Titles:
163 Nights
The Bean Pot and Other Tales
What About Bob?

How to Watch Movies and Television Shows For Free

Annie Jean Brewer

FIRST PRINT EDITION

Copyright 2012 by Annie Jean Brewer
Annienygma.com

PUBLISHED BY:
Annie Jean Brewer on CreateSpace

All rights reserved. No part of this publication may be reproduced, stored in a retrieval system or transmitted by any means, electronic, mechanical, photocopying or otherwise.

*Dedicated to everyone out there
who wants to save money.*

Contents

Introduction 1
Note 3
A Request 5
Disclaimer 7
A Few Assumptions 9
Protect Yourself 11
Software Needed 13
Use Your Television 17
Video Websites 19
Registration 29
Torrents and File Sharing 31
Popups and Ads 33
How to View Movies and Shows 35
Help! My Site is Gone! 39
Share the Love 43
Conclusion 45
About the Author 47

Introduction

We cut the cable cord almost a decade ago when, newly divorced, I simply could not afford to pay for the service. There wasn't a lot of stuff available in those days but we supplemented the videos we found with movies from libraries and borrowed from friends.

The number of websites offering movies and television shows online has skyrocketed since then. My daughter and I rarely pay to watch anything anymore because we don't have to. A quick search among our favorite websites or a visit to a search engine produces whatever we want to watch so easily that it reminds me of "shooting fish in a barrel."

For instance, if I hear of a movie that sounds interesting I open a browser, visit my favorite website and search for the title. If it is available there I scroll through the links available and pick my favorite. Within minutes I am watching the film.

It is the same for television shows, anime, documentaries and other types of videos.

If it isn't available on one site, chances are high that it will be available on another one. The only shows I've had trouble viewing are sporting events and movies that aren't fully released yet.

In this uncertain economy entertainment can mean the difference between having hope and feeling hopeless. Few things are more disheartening than telling your children that they can no longer watch their favorite shows because Mommy or Daddy lost their job.

This method allows you to keep the family happy and save money at the same time.

In this ebook you will learn how to prepare your computer, protect it from predators, find the shows and movies you want to watch and even how to connect your computer to certain televisions. You will discover rips, cams and why file-sharing is dangerous and should be avoided.

You will not only receive a list of popular movie and television websites, you will also learn how to find new sites as they become available using search engine tricks and how to determine if your favorite site is down—or if it is just you.

If you are ready to save hundreds, possibly thousands of dollars a year on entertainment buckle up—it's time to take a ride on the Internet!

Note

This book was written primarily for Windows computers. Mac users can use the links but the videos based on Flash may not work without an appropriate workaround.

A advantage of using this report with a Mac or a Linux computer would be the fact that you won't have to worry so much about viruses—a huge advantage.

Since the majority of computer users out there still run Windows, the software section of this report deals primarily with that operating system.

Also, while I have made every effort to provide accurate website addresses and other contact information at the time of publication I do not assume any responsibility for errors or changes that occur after publication. I have no control over any third-party websites and their content do not assume any responsibility over them whatsoever.

A Request

Please leave an honest review of this ebook at the website of purchase. This will help others like you determine if this book will help them. To thank you for your review I will be happy to send you a PDF copy of this book so that you can store it on your computer or print it out as desired. Please email me at annie@annienygma.com with a link to your published review and I will send you the PDF copy of this book.

Thank you!

Annienygma

Disclaimer

In this sue-happy age I've no choice but to offer a disclaimer:

I am not your mother. I am neither your attorney nor your computer tech. I am not aware of the laws in your specific area and make no claims to the legalities of the methods described within. All information contained within this ebook is for information purposes only offered on a "use at your own risk" basis. I will neither fix your computer nor bail you out of jail if anything negative results from trying the ideas within this book. Please use common sense and check the local laws in your area before using any of the tips within these pages. The laws may change over time and what is legal at the time of writing may become illegal in the future.

As the legal eagles like to say, before relying on this material users should carefully evaluate its accuracy, currency, completeness and relevance for their purposes and obtain any appropriate professional advice relevant to their particular circumstances.

In other words, use the information contained in this book at your own risk. While I personally use this stuff and have for years it may not be legal in your area by the time you read this book so check your local laws. I am not responsible for anything that happens to you or your computer if you try the methods in this book.

Now that we understand each other, let's go watch some movies!

A Few Assumptions

We have all heard what is said about making assumptions but in this case I need to make a few about you. Here they are:

1. You own a Windows-based computer running Windows XP or higher.
2. You have access to a high-speed internet connection.
3. You know the basics of using a computer and navigating the internet.
4. You enjoy watching movies and television shows.
5. You don't want to spend money to watch movies and television shows.
6. You are willing to learn something new.

If these apply to you then we're good to go. If not you may want to pass on this volume right now; we may not be a good fit for each other.

Protect Yourself

Whenever you visit unfamiliar places there is always a bit of risk. With a computer that means that you could become infected with spyware, viruses or other forms of malware. If you already have some malware installed on your computer there is a possibility that the malware could be triggered.

Before you start I recommend that you make sure your computer is protected with an up-to-date malware solution. Microsoft offers a free one called Microsoft Security Essentials which is available at the following link:

http://www.microsoft.com/security/pc-security/mse.aspx

After installing an antivirus program run a complete malware scan on your computer system and remove anything that is detected. If you have any issues you may want to check out my ebook Professional Help: How to Prevent and Fix Malware, Viruses, Spyware and Other

Baddies[1]. It is filled with instructions to help you repair a computer that is infected and also contains links to other free antivirus programs.

Another free program you can use to scan for malware is Spybot Search and Destroy[2]. This program is regularly updated and is an excellent alternative to commercial spyware scanners. I have used it along with other software to clean infected computers for years. If you like the program please consider donating a small amount to the creators to help them continue work on this program.

Whether you decide to watch movies online or not please make sure that your computer is free of malware. This will protect your privacy, improve the performance of your computer and vastly improve your computing experience.

---

[1] http://www.amazon.com/gp/product/B004GNFI7C/ref=as_l i_tf_tl?ie=UTF8&camp=1789&creative=9325&creativeASIN=B004 GNFI7C&linkCode=as2&tag=ajourn0c7-20

[2] http://www.safer-networking.org/index2.html

Software Needed

For the best experience when watching movies online I recommend that you use the Mozilla Firefox browser[3].

While you can watch movies and television shows online using almost any browser Firefox has two plugins that will help to improve your viewing experience. NoScript[4] is a Firefox extension designed to protect you from advertising, clickjacking and other browser hazards. To use, you enable the scripts you wish to run on each page you visit (you can permanently grant permission to sites you visit regularly). NoScript will then block unwanted ads and other scripts on the pages to protect you and provide a better browsing experience. This is the plugin I use to reduce the amount of ads I see when I watch videos online.

VideoDownloadHelper[5] is a plugin that will help immensely with choppy videos. On videos that are choppy, you can manually cache them to your hard drive using this

---

[3] You can download the most recent version at this link: http://www.mozilla.org/en-US/

[4] http://noscript.net/

[5] https://addons.mozilla.org/en-US/firefox/addon/video-downloadhelper/

plugin and watch them at your convenience after the download is complete. Simply delete the file when you are done.

VLC[6] is a media player that can play a large variety of the files that you will find online. Like Firefox, NoScript and VideoDownloadHelper you should not have to pay for the basic version of this software.

Warning: Some websites will try to persuade you to download and install their special software to view movies on their sites. **Do Not Do This**. In my experience these programs are filled with spyware and other nasties that can mess up your computer. One telltale sign is when a website tells you that you need to install VLC media player at a certain link—and you already have VLC media player installed on your computer. Avoid these websites and do NOT install any type of software on your computer that originates from them.

Some videos are only available in Flash format. Adobe has a free flash player available here: http://get.adobe.com/flashplayer/. I also recommend JW Player[7] for flash videos as well. The reason I recommend JW Player is that some websites will try to trick you into downloading a contaminated version of this player. You

---
[6] http://www.videolan.org/vlc/
[7] http://www.longtailvideo.com/players/jw-flv-player/

can avoid this by already having it installed and paying attention to what you are downloading. If you already have the program installed and a website claims that you do not you will know that they are attempting to install malware on your computer. Close that window and move to the next website.

Some websites may require that you have the Xvid codec. As with the other programs, do NOT download it from the website that asks for it—their link may be contaminated. Instead visit this link to download it direct: http://www.xvid.org/Downloads.15.0.html. This is not a very common codec so I don't recommend installing it unless you find that you need it.

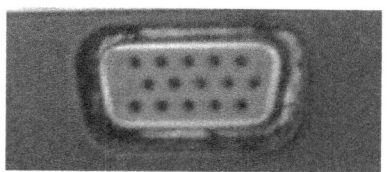

VGA port

Use Your Television

Some televisions have a VGA port. This port allows your television to be used as a monitor for your computer, allowing you to watch your movies on the big television screen instead of on your small computer screen. This is the most common port available at this time that will allow you to take advantage of the bigger screen on your television. To use this simply connect a VGA cable from the VGA out port on your computer (most laptops even have this port) to the VGA in port on your television. Configure your television to accept the input from the VGA port (follow the instructions in your television manual) and you will be able to use your television as a large computer monitor, greatly enhancing your viewing pleasure.

Video Websites

There are an incredible number of video websites available that offer both movies and television shows for your streaming pleasure. In fact, there are so many of these free sites available that you may never want to pay to watch a movie or a television show again.

For instance, I watched The Expendables 2 within days of the initial release. It was a cam (recorded with a video camera) but the quality was good and the film enjoyable. The price (free!) could not be beat.

After watching the movie Avengers I was intrigued by one of the characters so I did a quick search and found he had his own movie entitled Thor. Then I discovered that Chris Hemsworth (the actor who plays Thor) was also in the movie Cabin in the Woods; I ended up having a movie marathon without leaving my kitchen table.

While most new releases are poor-quality cams, occasionally you will discover a DVD quality video or even a movie that isn't released yet. I watched The Lovely Bones two weeks before it was released from an excellent DVD rip (converted from DVD format for online viewing).

I generally keep an eye out for new releases that interest me; in most cases I can watch them within a week of their release to theatres, though some take longer. If the cam that is available is of poor quality I may decide to wait until a better quality video comes out but sometimes that may not be until the movie is released to DVD. Depending upon the quality of the videos you find you may wish to do the same.

Television shows are a bit different than movies in the quality aspect. They are generally recorded directly off of the television so they are the equivalent of DVD quality from the start. Occasionally you will see little insets from the television channel and I've even seen snippets of commercials but it rarely distracts from the viewing experience. The best part of watching television shows online is the fact that you can have a marathon of your favorite show at any time of the day or night. It is like having an incredible video library at your disposal every day.

I have watched every episode of <u>The Big Bang Theory</u> and <u>Dexter</u> that is available but I have yet to pay for a single monthly subscription. I don't even know what station shows <u>The Big Bang Theory</u>—it was recommended by a friend and I searched for it online. These episodes are generally posted a few days after the paying customers get to watch them; sometimes I may have to wait up to a week for the latest episode. As a result I prefer to allow several episodes to "build up" that I haven't watched yet and have a marathon to get caught up on all of them at once, which greatly improves the enjoyment of the storyline.

Hulu only keeps a single season of The Pretender available for free viewing but I am currently watching Season 2 from another site. My kid enjoys watching a variety of anime shows like Peach Girl and others whose names I don't recall but she also enjoys Barbie movies, all for free.

Here is a list of movie and television show sites that I personally recommend:

Hulu[8]. This site has a lot of commercials. They also heavily promote their premium content but it is an excellent website to watch movies and television shows regardless. The site is easy to use, malware-free and streams easily. It is the easiest site to use when you are just getting started with online video. They have movies, television shows, outtakes, trailers and other videos in a format that is searchable or browsable, depending upon your preferences.

---

[8] http://www.hulu.com

YouTube[9] is one site with a huge variety of video. I have discovered whole movies (uploaded in segments), television shows, webisodes, special segments, music videos, documentary, newscasts and many other videos. The quality ranges from poor to excellent but the variety is insane. This is definitely one site to always keep on your list.

OVGuide[10]. This website provides links to a large number of movies, television shows and other videos. Type the name of the video you are looking for into the search box to receive a list of links for you to explore. This site has been around for ages and is an excellent place to start when you are looking for an obscure movie.

Wally TV[11] is another website you may wish to check out. The site is clean and they feature videos from a number of legitimate sites and stations.

AnimeFreak[12] and WatchCartoonOnline[13] are the two main anime sites that my daughter uses. She loves the volume of anime available on them.

1Channel[14] is my absolute favorite site for watching online video. The site carries links to both movies and

---

[9] http://www.youtube.com
[10] http://www.ovguide.com/
[11] http://www.wally.tv
[12] http://www.animefreak.tv/glossary
[13] http://www.watchcartoononline.com/

television shows and you can browse by a variety of methods. The two most common browsing views are "recently added" and "featured." The "featured" section will show the most popular movies (or television shows, depending upon what you are looking for) in order of the most popular. You can find almost anything on that site, which will display a list of links to the video with an icon indicating quality and information about the number of times downloaded, ratings and reviews by members of the website. Unlike many video websites, registration at 1Channel is free and gives you the ability to report broken links and rate videos on the site. To watch a video you will need to click on one of the links in the list to be taken to their website. Putlocker and Sockshare links are among the best available on this site.

[14] http://1channel.ch

Justin TV[15] is a website where you can watch a variety of channels that stream complete collections of movies, television series, vanity cams, documentaries and other items. These channels stream live so it is very similar to watching traditional television. I know of several people who can spend all day on this site without ever leaving and consider this their new television. One of the wonderful aspects of Justin TV is that you can engage in live chats with other users, moderators and the hosts of other channels. If you develop a friendship with the channel hosts you can actually request for them to stream particular shows or movies. This is a very fun site to hang out in and is quite addictive.

---

[15] http://www.justin.tv

Here is a copy of my personal "cheat sheet" of video website links:

www.divxstage.net. Sign up for a free account and you can stream movies for free.

http://www.letmewatchthis.com

http://www1.zmovie.tv

http://zmovie.eu

http://79.142.69.71

http://www.novamov.com

http://www.veoh.com

http://joox.com

http://motionempire.com/Watch_Free_Movies.html

http://www.movie2k.to

http://www.crackle.com

http://www.moviesplanet.com

http://www.youku.com Chinese language website; if you visit it using the Translate This extension in FireFox (https://github.com/BoringCode/Translate-This---Firefox-Addon), the Google Chrome internet browser (https://www.google.com/intl/en/chrome/browser/?hl=en&brand=CHFX&utm_campaign=en&utm_source=en-oa-na-us-bk-bng&utm_medium=oa) or you can also right-click an open spot on the page in Internet Explorer and select to translate with Bing.

http://www.tudou.com/ (Chinese language website; translate using the same tips as above)

http://asian-horror-movies.com/ (only Asian horror movies)

http://www.dailymotion.com/us

http://www.24world.tv

http://alluc.org

http://www.chooseandwatch.com

http://www.coolstreaming.us

http://www.tvduck.com

http://www.ustream.tv/

http://www.metacafe.com/

www.craftytv.com

http://www.divxcrawler.org/

http://www.free-tv-video-online.me/

http://www.freetvsearch.com/

http://vimeo.com/

http://current.com/

http://www.ted.com/

http://bigthink.com/

http://www.comedycentral.com/cc-studios

http://blip.tv/

http://www.5min.com/

http://wwitv.com/

http://www.stickam.com/

How to Watch Movies and Television Shows For Free

http://www.blinkx.com/

http://www.channelchooser.com

http://channelchooser.com

http://internettvlist.com

http://wwitv.com/portal.htm

http://www.live-online-tv.com/

http://www.lordoftv.com/

http://www.myturn.tvheaven.com/

http://www.piggymoo.com/

http://www.sidereel.com/

http://www.streamick.com/index.php

http://www.watchmoviez.com/

http://www.yourtvlinks.com/

http://www.alluc.org/

http://allsp.ch/

http://www.channelchooser.com/

http://tv-video.net/

http://streamingtvepisodes.com/

http://online-television.tv/

http://moviesdesk.com/

http://movies.nabolister.com/

http://www.tv-links.cc/

http://watch0nline.info/

http://tvunderground.org.ru/

http://www.cheri-movie.com/

This is not a complete list of movie and television streaming websites. New places come and go almost daily so if one link goes down you can always find another one to replace it. All of these links are live as of this writing; how long they will stay live with the transient nature of the Internet is unknown. I recommend that you keep a list of video websites that you discover during your journeys; this way when one of your favorite sites goes down you will have several alternatives to choose from with minimum interruption of your viewing experience.

Registration

Some free video websites ask that you register for free before watching their movies. Personally I do not feel safe doing this but I know several people who register routinely. They do not use their real name or email address for safety, instead they create a disposable email account on one of the free services like Yahoo or Gmail.

Other websites request money for ad-free viewing. I don't recommend giving out your credit card information **at all** and seriously hope that you don't do this for safety reasons. While I have not heard of anyone getting ripped off by one of these sites the transient nature of them could mean that the site disappears shortly after you give them your money. If you want to pay money to watch videos online go to a well-known site like Netflix, Hulu Plus or Amazon Prime.

Torrents and File Sharing

Torrents are one way that some people are downloading movies from the Internet. This involves a special program called a BitTorrent client.

While using torrent software in itself is not illegal, uploading and sharing copyrighted content **is** illegal in many areas of the world. BitTorrent clients not only act as downloaders but they also share the downloaded pieces of the files so that while you are downloading the latest Hunger Games movie you are also uploading it to others as well and it is the uploading part that can get you into some serious legal trouble.

Watchdogs have been hired by various interests to monitor certain copyrighted content that has been made available on the Torrent networks. When your IP address is flagged as downloading the file they also log who your computer is sharing that file with. This information can be used to levy extreme penalties for file sharing.

The only way currently known to safely avoid detection while torrenting is to use a computer or virtual machine that does not contain any of your personal information, turn off IPV6 on that system and use a trustworthy non-US-based VPN client to route your BitTorrent traffic through. This method, depending upon the VPN service you use can get rather expensive and is not without risks because of the constant efforts by the watchdogs to overcome this hurdle. It would be much less expensive and risky to just use a legal service to watch your videos instead.

These same watchdogs also watch the file sharing programs out there (Limewire, Kazaa, etc.). While in a lot of cases they will just contact your Internet service provider to send you a threatening cease and desist letter the first time you are caught, you may not be one of the lucky ones. Avoid file sharing and torrenting for maximum safety and make sure that your kids don't do it behind your back[16].

Note that viewing streamed files is currently viewed in a different legal light than actually sharing them. Currently the watchdogs in the legal system are focusing on the ones who are sharing the movies online instead of the viewers.

Be safe. Don't share or upload copyrighted video files.

---

[16]

http://www.telegraph.co.uk/technology/3340148/Parents-of-net-music-thieves-to-be-fined-4000.html

Popups and Ads

While watching movies online at free websites you may be bombarded with popups. This is a normal part of the process. These websites pay their bills with the income from these ads; without them we would have to pay for the opportunity to watch stuff online. They are annoying at times but it just goes with the territory.

Check the taskbar at the bottom of your computer screen for any extra windows that the video site may have opened when you start watching your video. Close every window except the one that your movie is playing in. This will maximize your viewing experience; some of these windows will play noisy ads that will detract from your movie.

If you hear something distracting while your video plays check for windows that have been quietly opened in the background. Close these windows and the noise should cease.

Occasionally you will hear an advertisement playing even after you close all of those windows. Check the ads near the movie window and turn off the sound or use

NoScript to block them. You can also try maximizing the video; some websites turn off the sound on ads automatically when the video window is maximized.

How to View Movies and Shows

Depending upon the website and your internet connection the video you are trying to watch may end up being choppy. There are a few things you can do to correct this.

Internet Connection

Ensure that your Internet connection is fast enough to handle streaming video. Dial-up connections are incapable of streaming most videos and some wireless (cellular) and satellite connections may not provide the bandwidth for regular streaming. DSL, fiber or cable internet connections are currently the best options for those who wish to stream online video with any regularity.

If just one or two people in your household are streaming videos you can (in my experience) successfully use the slower DSL and cable internet packages. If you have several people watching movies online at the same time, use your internet connection as a VOIP phone service or otherwise experience quality issues you may wish to consider upgrading your Internet service to a faster package or try a lower number of simultaneous video streams.

Buffering

Buffering is the process used by computers to actually play online video. The video file is transferred to your computer's cache (temporary storage area) for playback. If the playback is faster than the video is downloading then the video will stop until the download catches up.

If you suspect that your computer is playing the video faster than you are receiving it pause the video and wait for a few minutes. This will allow the computer to download enough of the video so that you can watch without interruption. On some sites I start a movie and then pause it to wash dishes, do laundry or run an errand so that it has plenty of time to buffer before I'm ready to sit down and watch it. Note that if the browser window is closed you will have to completely start over with the buffering process—the computer deletes the temporary file as soon as you are finished to save on file space.

Manual Caching

To avoid losing a buffered video due to a browser crash, power outage or other mishap you may prefer to manually cache your videos when the playback is choppy.

How to Watch Movies and Television Shows For Free

When you manually cache the video you deliberately save it to a designated spot on your computer so that you can watch it at your convenience without buffering concerns. This is also an excellent option for those movies you wish to watch more than once.

To manually cache a movie open the video website using Firefox and the VideoDownloadHelper extension start the video and wait for the three color dots on your browser bar to activate. Click on those dots and instruct VideoDownloadHelper to save the video file to your hard drive. Designate a spot (the Desktop is a convenient location) where the files are saved automatically to save time. Firefox will then download the video completely. You can choose to watch the video on the website or wait until the download has finished.

Not all websites are compatible with VideoDownloadHelper. Some will appear to download the video but will only provide a small file. Don't stress over this; just delete the file and try with another website.

You can also manually cache some videos in Internet Explorer if you install the Freecorder toolbar[17]. This toolbar, while quite handy, is known to run adware while it is active. Disable this toolbar and restart Internet Explorer when you aren't using it to avoid the annoying popups.

---

[17] http://www.freecorder.com

Help! My Site is Gone!

Sometimes governmental powers will decide to shut down[18] our favorite websites. When this happens the site will either disappear or you may see a page that looks like this:

This domain name has been seized by ICE - Homeland Security Investigations, pursuant to a seizure warrant issued by a United States District Court under the authority of 18 U.S.C. §§ 981 and 2323.

Willful copyright infringement is a federal crime that carries penalties for first time offenders of up to five years in federal prison, a $250,000 fine, forfeiture and restitution (17 U.S.C § 506, 18 U.S.C. § 2319). Intentionally and knowingly trafficking in counterfeit goods is a federal crime that carries penalties for first time offenders of up to ten years in federal prison, a $2,000,000 fine, forfeiture and restitution (18 U.S.C. § 2320).

. Sometimes the ones operating the site will decide to close up shop and move to other projects. This is a fact of life with video websites so you need to know how to deal with this inevitability.

---

[18] http://www.nytimes.com/2010/11/27/technology/27torrent.html?_r=3&

Note: If you think that the issue may be with you and not with the website you can verify by visiting http://www.downforeveryoneorjustme.com. This website will tell you if the site is really down or if the problem is on your end. Of course, if you can't even open that window then you know the problem is yours.

Government takedowns are the main reason why I don't recommend registering at video websites. Who knows what the government will do with all of the information they gather as they seize these machines? Play it safe and keep your info to yourself or use fake information if you decide to take the risk. An IP address is a bit more anonymous and you can try to blame an unsecured or hacked router if they unexpectedly change the laws in your area.

When the eventual missing website happens to you open a browser and go to http://www.google.com, http://www.google.se or another search engine like http://www.dogpile.com and type in:

Watch ~movies online
watch free ~movies
~movie websites

These phrases and others will help to locate links that still work. The tilde is a search command that instructs search engines to also display similar content like videos,

TV and other items. Sometimes you can type in the name of your favorite website to search forum posts for the new location.

For instance, when my favorite website went down I searched for "What happened to Letmewatchthis?" in Google. This is the answer that appeared in the results: http://answers.yahoo.com/question/index?qid=20110910133722AA5C2WL.

That provided me with the new location of the website and I was able to start watching at the new location immediately.

It is unknown whether or not the problem of disappearing websites will ever go away so the ability to locate more is a must. Fortunately, modern search engines make them easy to find.

Share the Love

Quietly share your love of watching free movies and television online with your friends. This will increase the demand for this resource and encourage more websites to appear.

Networking will allow you to swap website information to maximize your viewing enjoyment. Don't neglect this step—I have found some of my favorite websites this way.

The best part of watching movies and television shows online is the fact that not only are you able to save money but you can teach your friends how to do so as well.

Conclusion

It is my hope that this report helps others save money by changing their viewing habits while encouraging media providers to offer their content at more reasonable rates.

If one person learns to save money using the tips in this book I will be content.

When you have finished with this ebook please help others by leaving an honest review on the website where you purchased this book. This helps enable others to determine whether or not this information will help them. Thank you.

About the Author

Annie Brewer is a frugal living expert who combines minimalism with frugality to be a stay at home single mom to her daughter. She is the author of the popular book The Shoestring Girl: How I Live on Practically Nothing And YOU Can Too!, The Minimalist Cleaning Method Expanded Edition and a number of other titles. You can learn more about her at Annienygma.com.

Connect with Annie Online

Website
http://annienygma.com

Amazon Author Central:
http://amazon.com/author/annienygma

Email
annie@annienygma.com

Facebook
http://www.facebook.com/annienygma

Smashwords
http://www.smashwords.com/profile/view/annienygma

Twitter
http://www.twitter.com/annienygma

Yahoo! Contributor Network
http://contributor.yahoo.com/user/annienygma

THE SHOESTRING GIRL

How I Live on
Practically Nothing…
And YOU Can Too!

ANNIE BREWER

Do You Want to Live on Less?

Would you like to learn how from someone who actually does?

Over ten years ago I found myself a single mother with three children to raise.

I had to learn fast.

I had to support those kids on a fast food paycheck while I put myself through school.

Not only did I manage to do it but I topped my own expectations. We ended up living better than I <u>ever</u> would have imagined.

Since then I have not only quit my day job but I have built up sufficient income to become a single stay-at-home mother to my youngest child. This feat would not have been possible without the frugality of shoestring living.

We live well on about $500 a month - and know how to live on even LESS!

Over the years I have shared my secrets with others who have fallen on hard times. I have helped friends who became disabled, single parents, the unemployed and others who found a need to live on as little money as possible.

The first thing I always shared was the timeless words of my grandmother. Even now I can hear her reminding me to hold up my head because...

Annie Jean Brewer

"There's no sin in being poor!"

This may be your first brush with life below the poverty line. You may be scared. You may be ashamed. You may not know what to do or where to start.

I'm here to help you save money

I have drawn upon my 10+ years of personal experience to create the ultimate frugal living guide. I won't bore you with stupid fluff about clipping coupons. Instead, you will find a concise method you can implement to save thousands of dollars over the course of a year.

Sections Include:
Housing
Auto
Groceries (Includes raising food)
Computers (includes where to find free and inexpensive software)
Television (includes watching shows online for free)
Books (lots of links to free ebooks and how to search for free ebooks online)
Music (includes links for free music sites)
Clothing
Cleaning tips and recipes
Personal care tips and recipes
Furniture
Thrift Shops
Yard Sales
Jobs and self-employment
And much more!

I not only explain the exact methods that I use to save money and live frugally but I also explain how I could live on about <u>half</u> of the money that I actually do.

While you may not wish to apply everything here I am confident that you will be inspired to save more money than you ever thought possible. You will learn the skills you need to overcome your current financial challenge.

Start Saving Money Today!

Available in both print and ebook format at many popular retailers.

HOW TO WRITE EBOOKS FOR A LIVING

By

Annie Brewer

Do you Want to Earn a Living from Ebooks?

As a single mother I asked the question: *How can I stay at home with my child but still pay the bills*?

Job after job kept taking me away from my daughter's fleeting childhood. My frustration grew every time I missed another milestone in her life.

I combed the Internet in search of the answer. I found several places online where you could work from home but many of these kept me literally chained to a computer for hours on end. There had to be a better way!

One day I stumbled upon a blogger selling ebooks from his website. Not only selling them, he was actually earning his living from ebook sales!

"I can do that!" I thought.

I contacted him, buttered him up and picked his brain.

Gleefully following his instructions I finished my first ebook, published it online and drooled at the screen in anticipation.

I sat, I watched, I waited. After my first few sales the money dried up like a puddle in the desert.

What was I doing wrong?

I went out in search of more writers and picked a few more brains. I stayed up late at night researching and experimenting, determined to become a successful ebook writer. I refused to give up and quit.

I discovered the secret to ebook success.

Now I spend my days at home instead of at the dreaded day job. I take long walks with my daughter instead of punching a time clock. Money comes automatically now so I can relax and enjoy my life.

Anyone can make a living with ebooks, GUARANTEED

If you follow the steps in this guide you are **guaranteed** to earn money with ebooks. I am so convinced that you will be able to earn a living entirely from ebook sales that I offer you a **6-month money-back guarantee.** If after 6 months of applying this method you are not earning money from your ebooks send me a copy of your purchase receipt and I will refund your purchase price.

How to Watch Movies and Television Shows For Free

This guide teaches you:
- What equipment you need to write ebooks
- What bank accounts you need
- How to financially prepare to live off your ebook royalty income
- Where to find the time to write
- The importance of a blog
- Where to practice writing in preparation
- Where to find subjects to write about
- How to create your ebook
- Where and how to create an ebook cover
- Ebook descriptions
- Where to distribute your ebook
- Ebook pricing
- The importance of a backlist
- Social media
- Making the leap by quitting your day job
- *And more!*

"A journey of a thousand miles begins with a single step." - Confucius.

Will you take that step today?

Where to Work Online

By Annie Jean Brewer

Do you want to work at home?

There are so many scams out there it is hard to determine legitimate work at home jobs. It took me years of searching and I stumbled upon my first legitimate opportunity entirely by chance.

Since then I have learned how to work entirely from home and have compiled a list of legitimate work at home opportunities. There is a little here for everyone as well as tips to avoid getting ripped off by the scams out there.

This book shows you:

- The Golden Rule to working online
- Money Matters
- Multiple Income Streams
- Fast Cash
- Tinkering Cash
- Searching Cash
- Writing Cash
- Therapy Cash
- Affiliate Links
- Roll Your Own (ebook that is..)
- Phone Actresses
- "Official" Jobs
- Clearinghouses
- The Big List of Online Jobs
- *And more!*

If you are serious about working online, this is the only book you need.

Annie Jean Brewer

"He that can live sparingly need not be rich."
Benjamin Franklin

There are a lot of frugality books out there.

I know. I've bought most of them.

Saving money isn't just a hobby for me; it is a way of life. It is what allows me to be a single stay-at-home mother for my child. We currently live on about $500 a month but if we wanted to we could easily live on **less**.

Here are just a few of the tips that I personally use to save thousands of dollars a year:

Tip #1 - Auto purchases. Annual Savings: $5,364.
Tip #32 - General Cleaning. Annual Savings: $216.
Tip #45 - Carpet cleaning. Annual Savings: $100.
Tip #67 - Salvaging stained clothing. Annual Savings: $50.
Tip #78 - Printer ink. Annual Savings: $100.
Tip #89 - Software. Annual Savings: $200
Tip #95 - Movies. Annual Savings: $52
Tip #100 - Television. Annual Savings: $1,200
Tip #114 - Credit Cards. Annual Savings: $480
Tip #129 - Where to work for Maximum savings. Annual savings: $1,011
Tip #241 - Housing. Annual Savings: $3,600

What could you do with that much extra money?

Annie Jean Brewer

Written by the author of <u>The Shoestring Girl: How I Live on Practically Nothing and You Can Too</u>, this guide covers:

- Auto
- Cleaning
- Computers
- Entertainment
- Finance
- Food
- Gardening
- General Household
- Housing
- Kids
- Personal Care
- Pets
- Shopping
- Travel
- Utilities
- Funeral expenses
- *And more*

Minus the fluff, this nitty-gritty guide immediately gets down to the business of saving money with *over 400 unique tips* designed to help anyone with a desire to save money.

You may not choose to use all of the frugal ideas in this guide but I am confident that this book will inspire you to **save more money** than you ever thought possible.

How Much Can YOU save?

How to Watch Movies and Television Shows For Free

Annie has not paid for a cable subscription in over a decade. Instead her family watches videos online for free. In this book she shares her tips, tricks and online wisdom to teach others how they can do the same.

This book covers:
- How to protect your computer before you start
- What software you need
- A list of video websites
- How to search for more websites
- What to do when your favorite site disappears
- Video viewing tips
- How to deal with Popups and other ads
- How to buffer videos
- How to manually cache videos
- Website registration cautions
- Why not to pay for using these sites
- Torrents
- File Sharing programs
- *And more!*

    Readers will not only have a resource of links to get started with but will learn how to discover even more viewing opportunities online and how to maximize their video experience while saving money in the process.

CONGRATULATIONS YOU HAVE REACHED THE END!
Thank you for your support!

***

Please help others—review this book.

www.ingramcontent.com/pod-product-compliance
Lightning Source LLC
Chambersburg PA
CBHW061516180526
45171CB00001B/206